MAZES
FOR KIDS 4-8

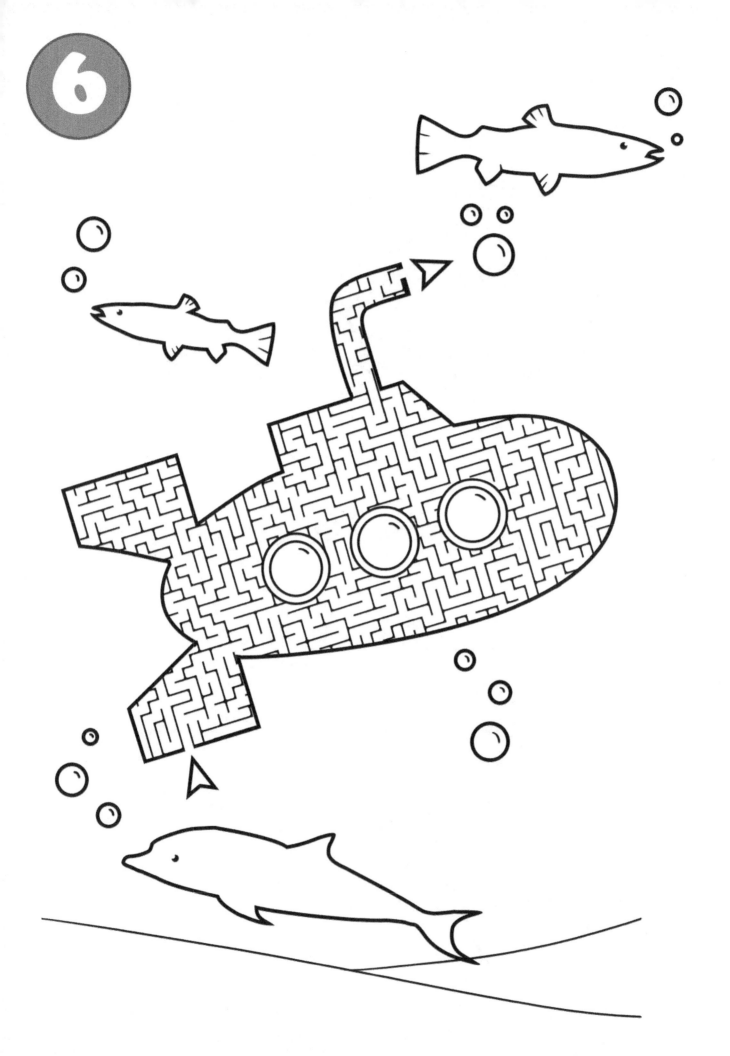

Tip: The ladybug must find the way to her flower.

14

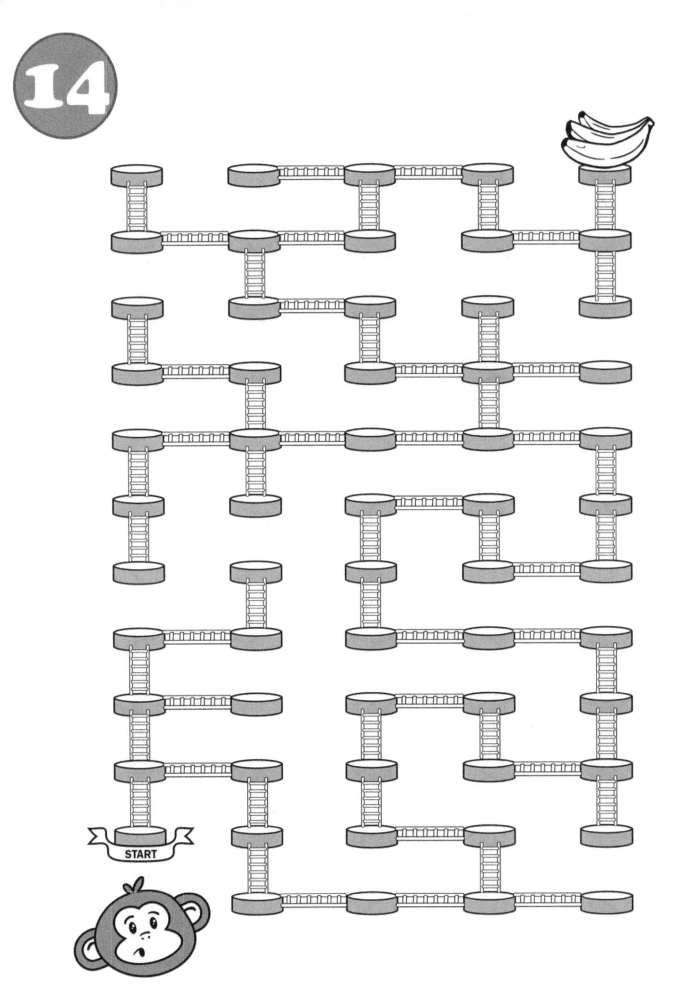

START

Tip: The monkey must climb the ladders to get to his bananas!

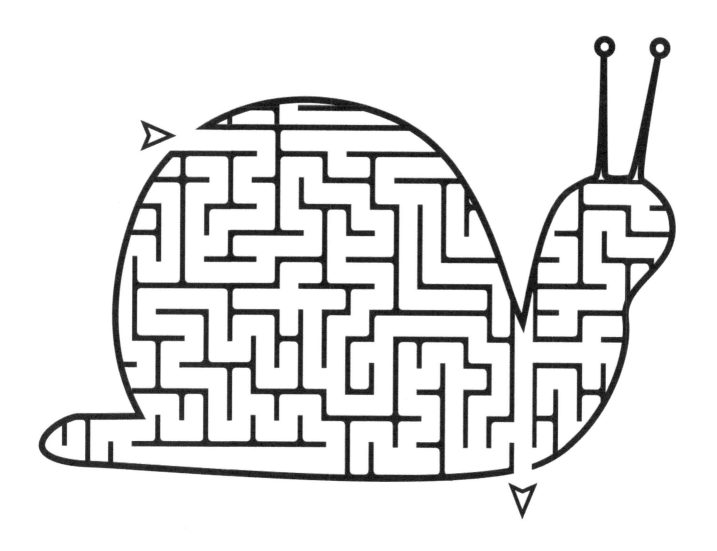

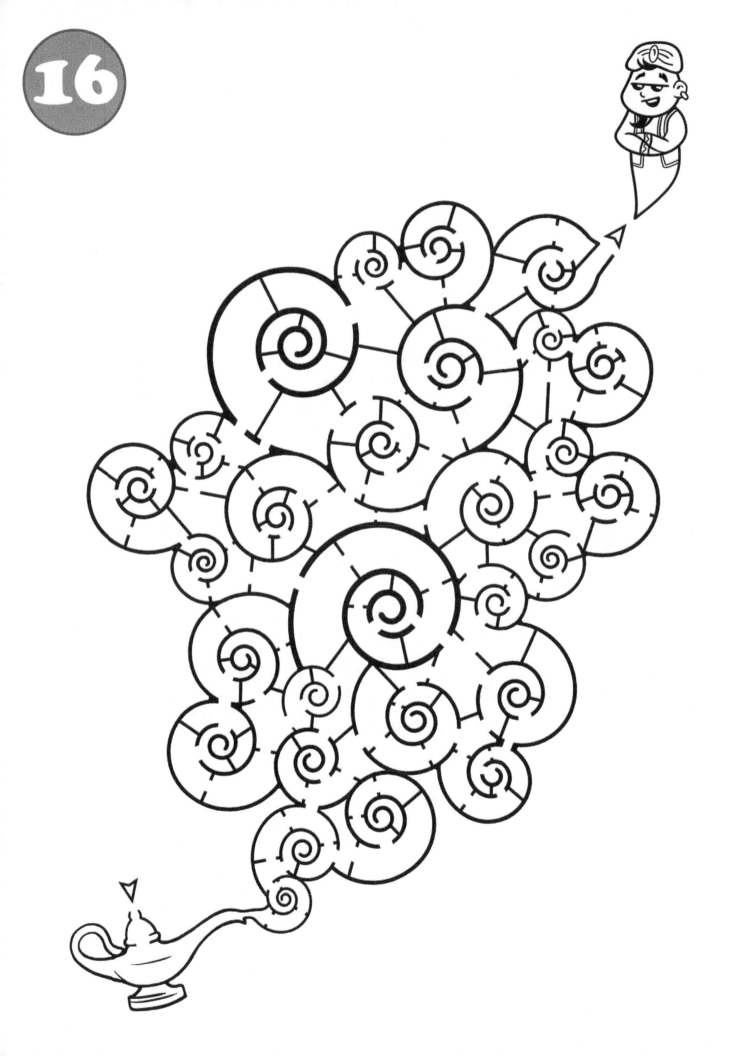

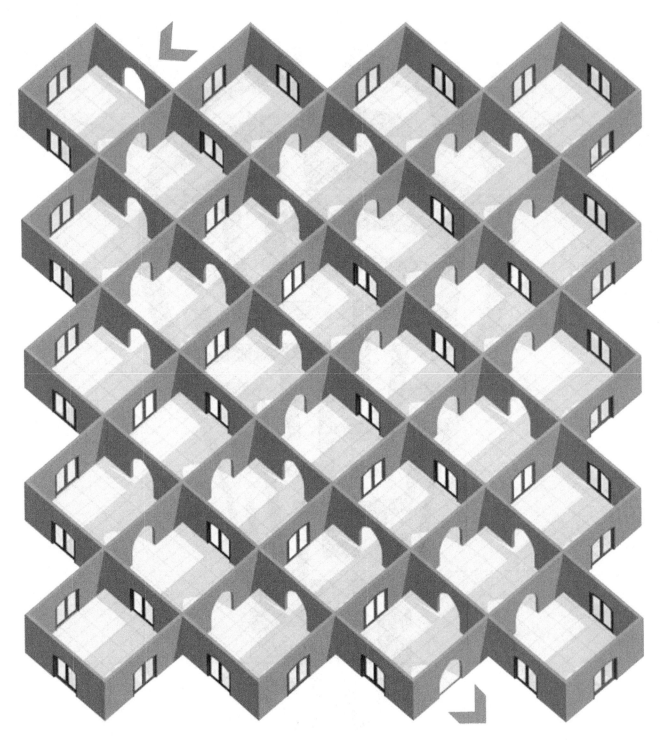

Tip: Find your way through the building without opening any windows!

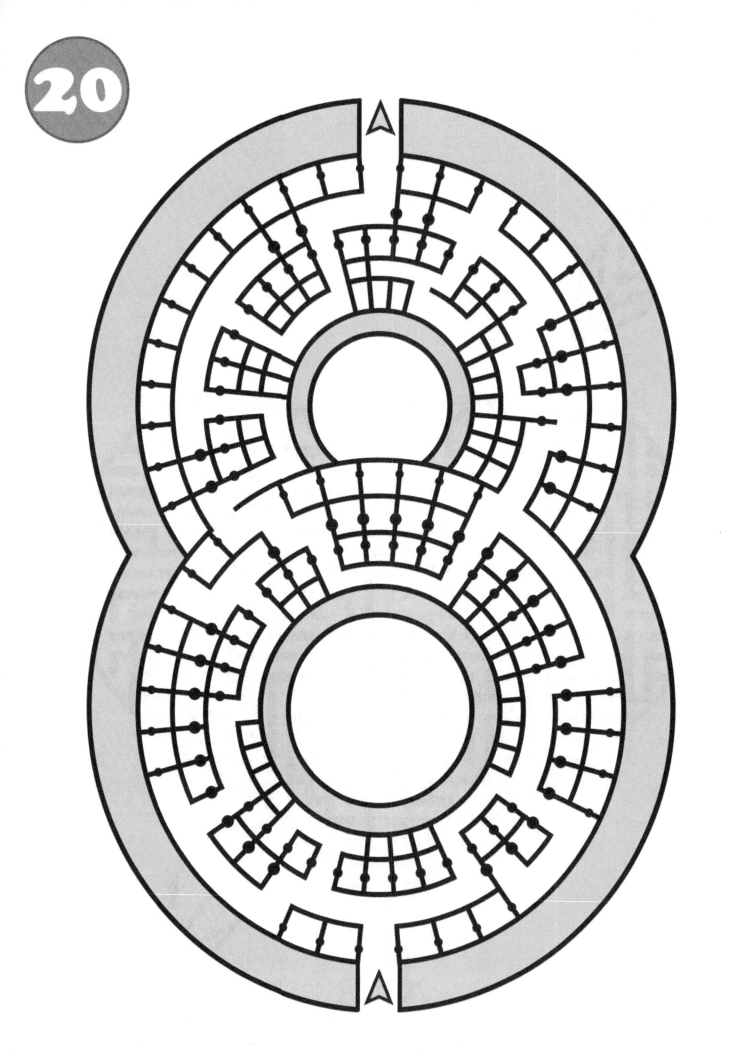

Tip: Walk along the top of the wall to reach the end!

Tip: Drive the car to the flag!

**Tip: The dog's paws can't touch the grass to get to his bone!
Stay on the stones!**

35

Tip: Drive the car to the trophy!

42

Tip: Each panda is connected to a different piece of fruit!

Finish

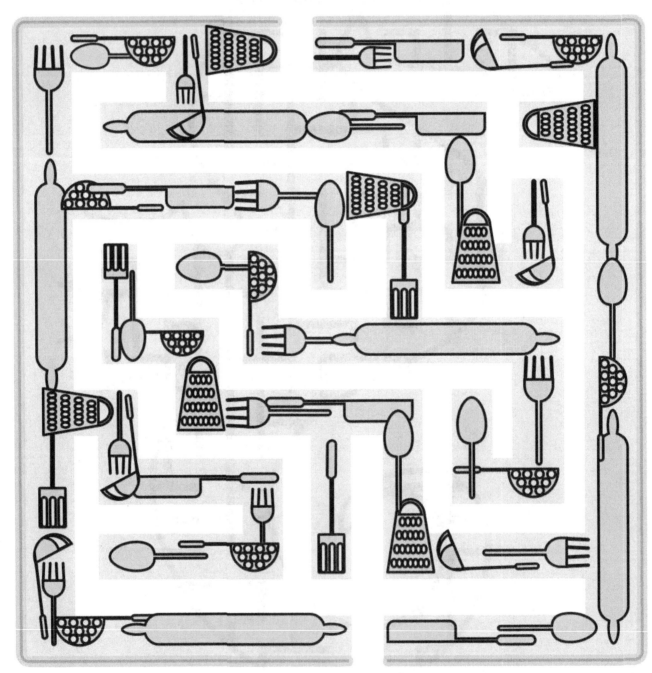

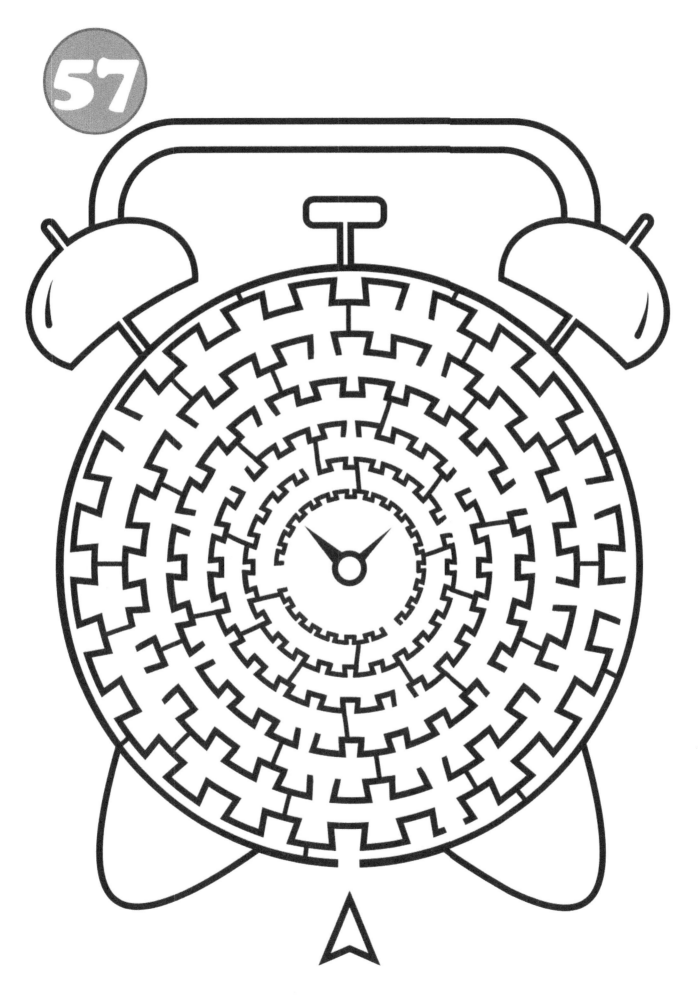

Tip: Try to get to the center of the clock!

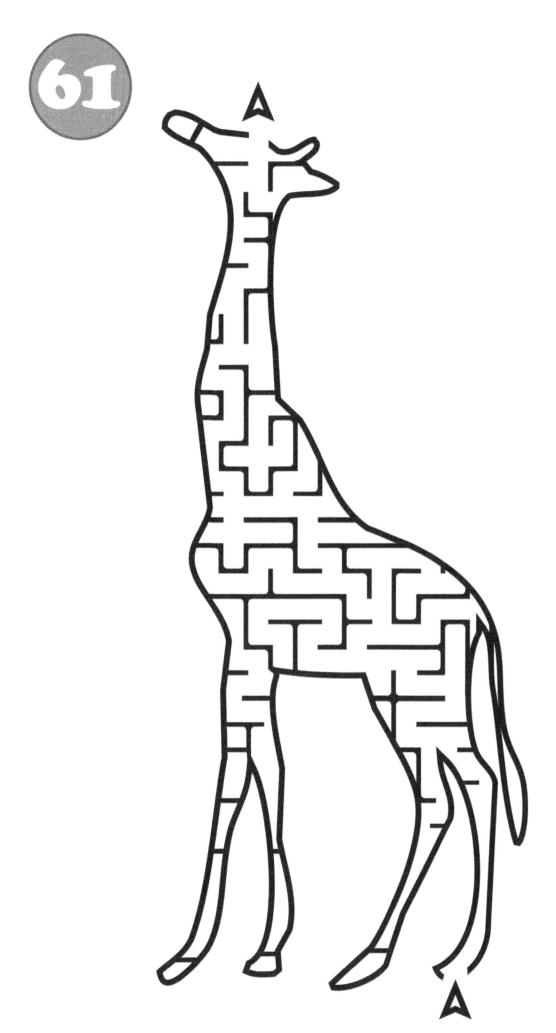

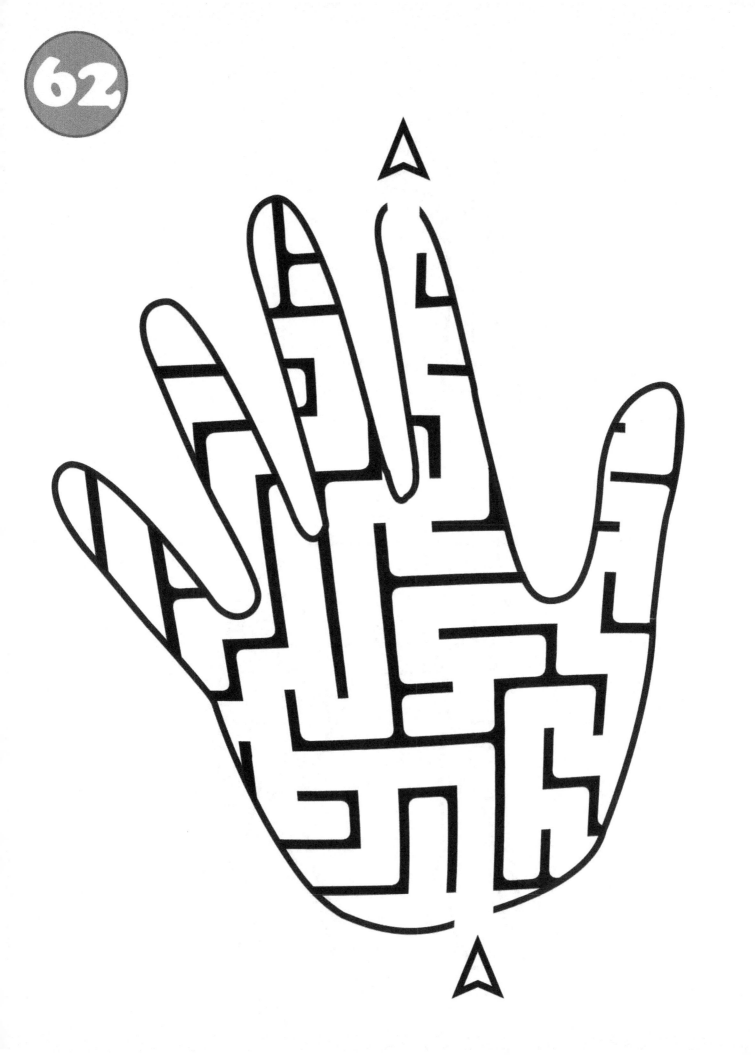

Tip: The squirrel must get to its acorn.

64

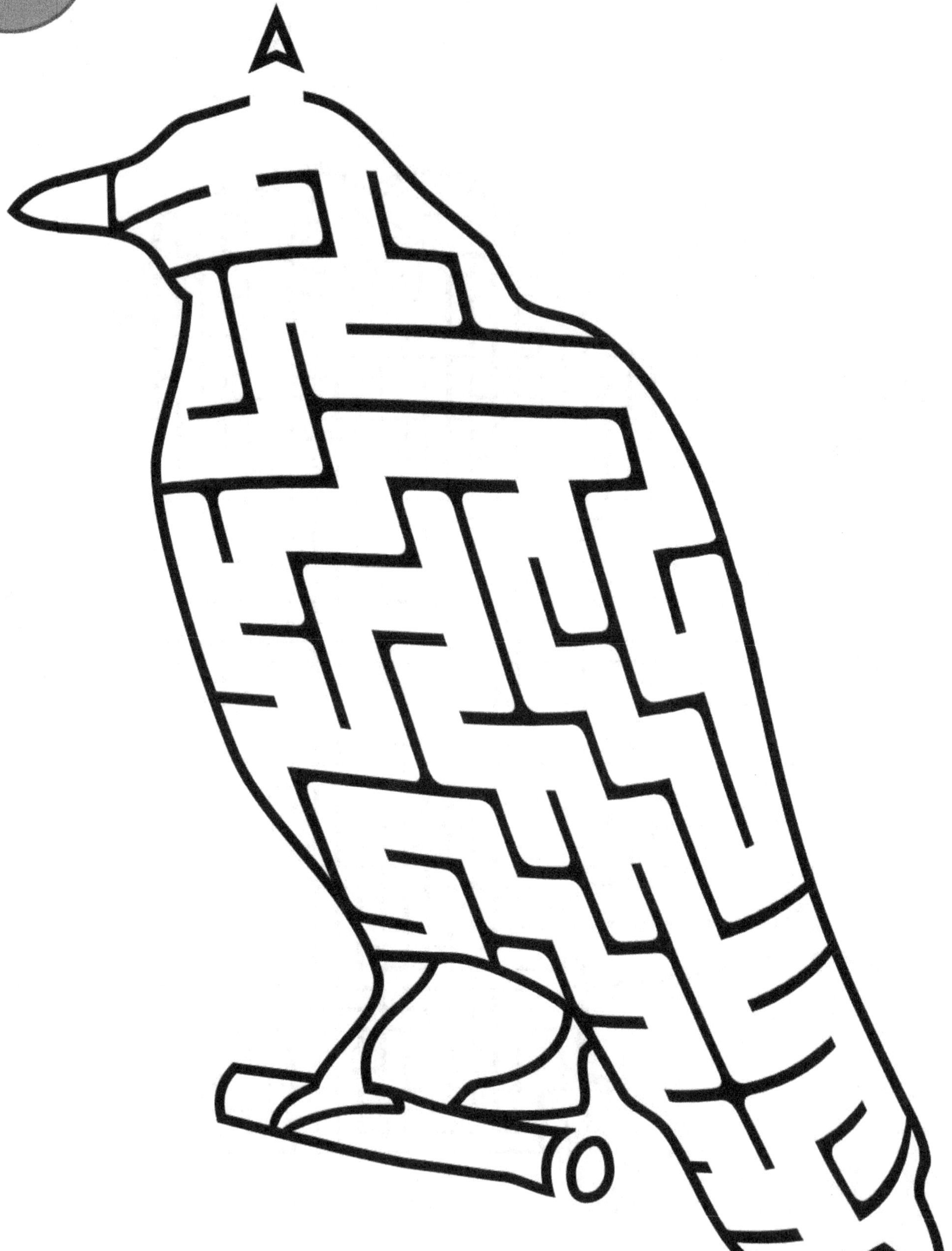

70

Tip: The mother whale needs to find the way to her baby.

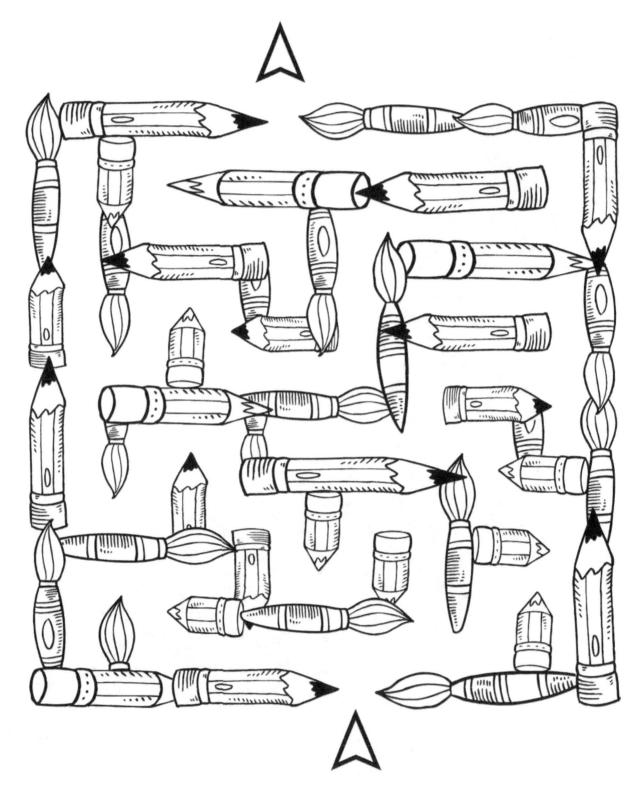

**Tip: The gingerbread man needs to
find the way home.**

**Tip: Connect each animal
to their tracks.**

Bear   **Duck**   **Dog**

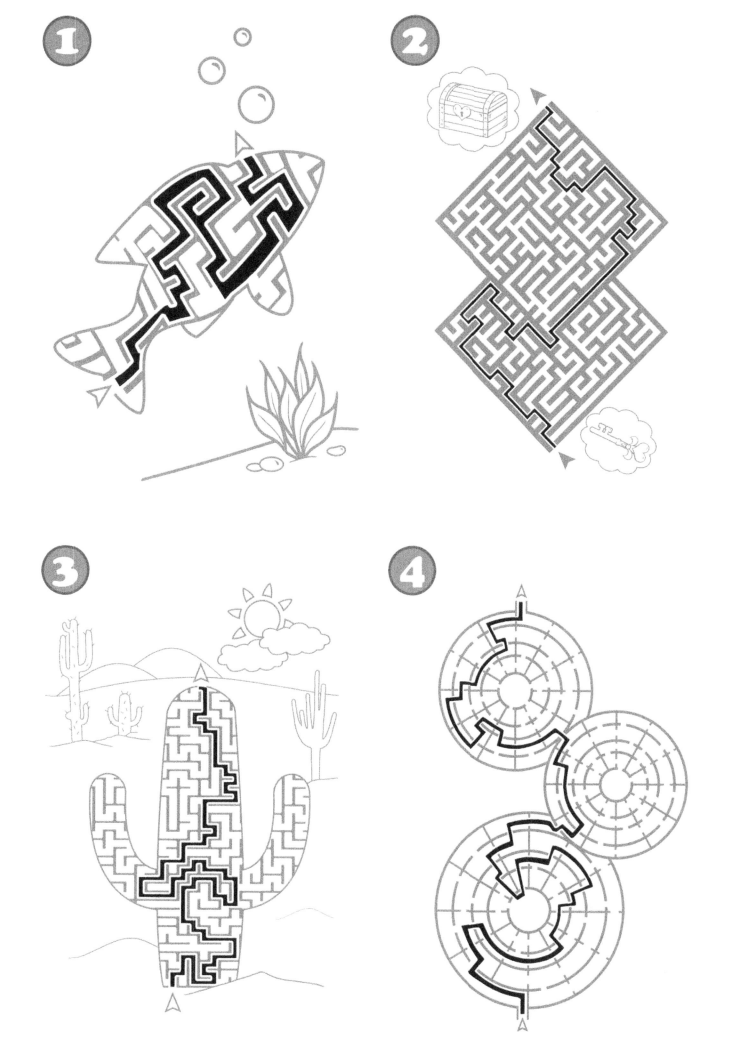

9

Tip: The ladybug must find the way to her flower.

10

11

12

13

14

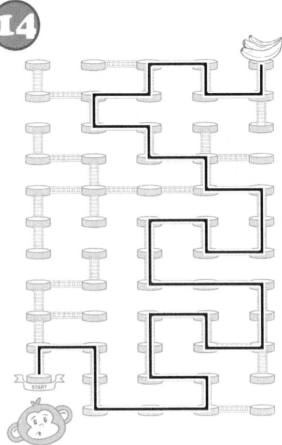

Tip: The monkey must climb the ladders to get to his bananas!

15

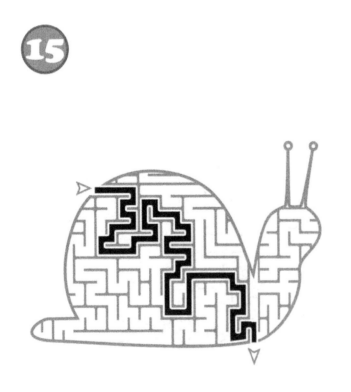

16

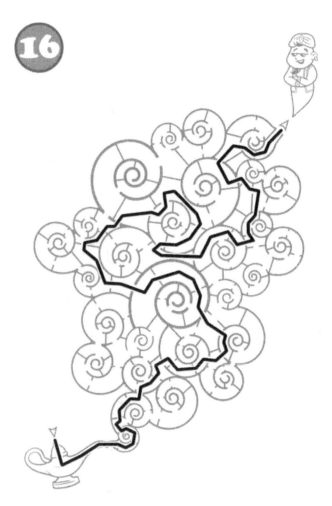

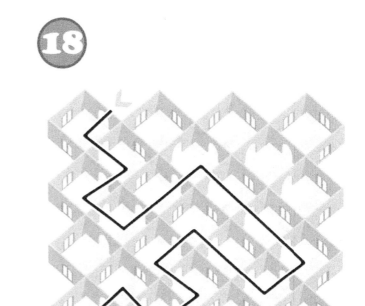

Tip: Find your way through the building without opening any windows!

Tip: Walk along the top of the wall to reach the end!

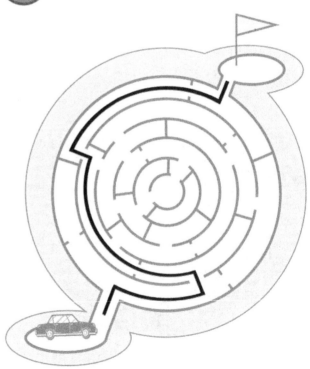

Tip: Drive the car to the flag!

25

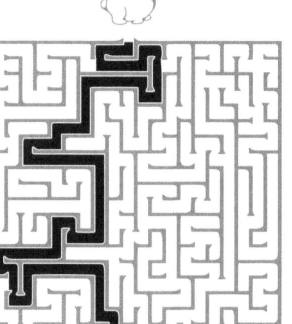

26

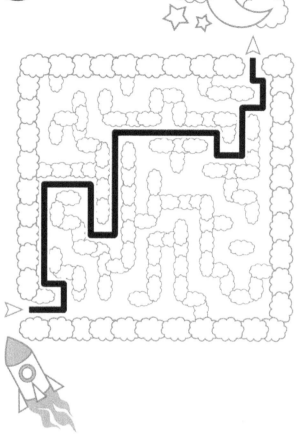

27

28

29

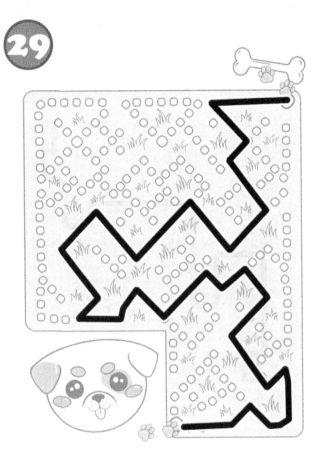

Tip: The dog's paws can't touch the grass to get to his bone!
Stay on the stones!

30

31

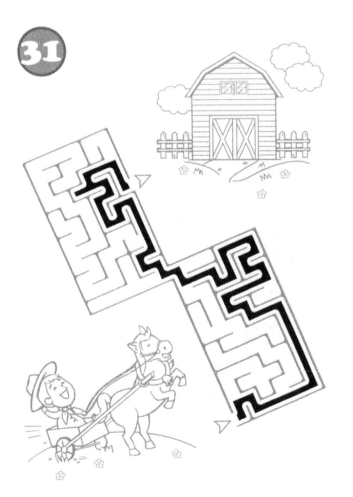

32

33

34

35

Tip: Drive the car to the trophy!

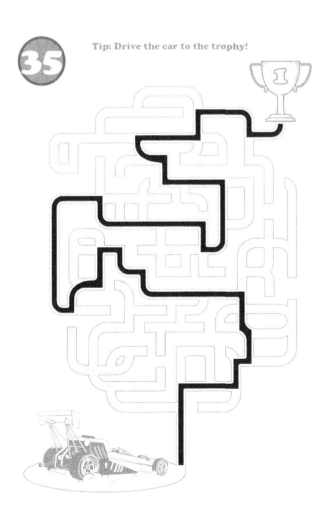

36

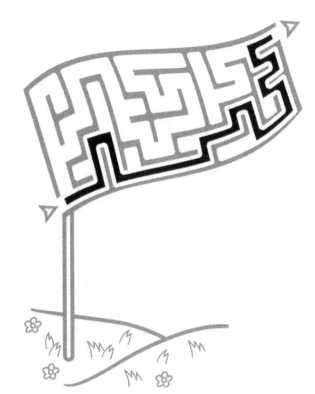

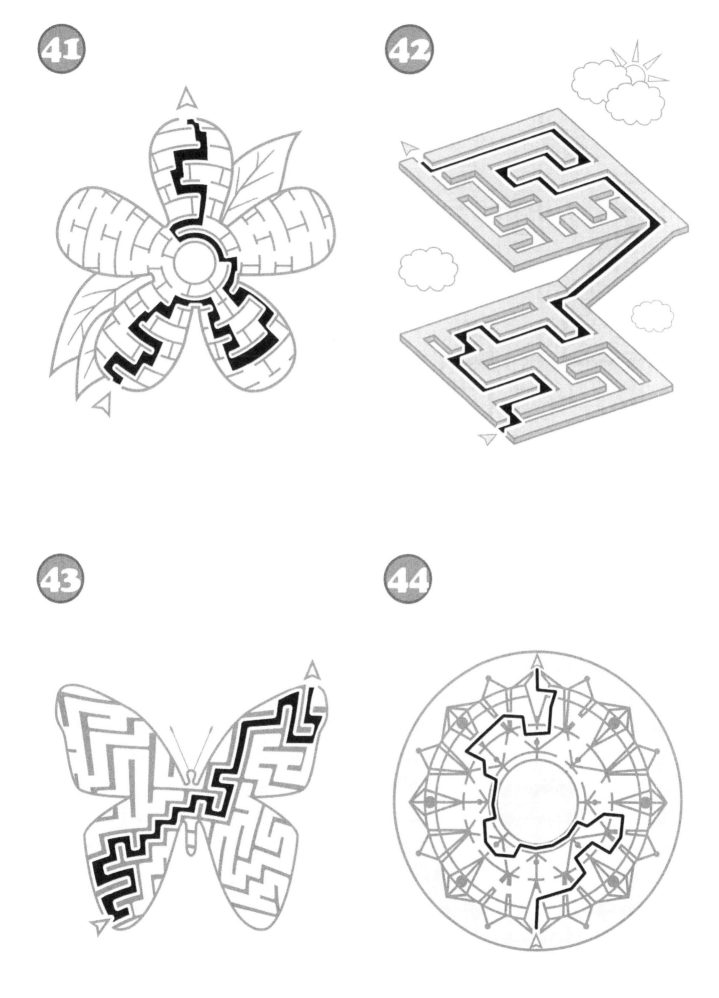

45

46

Tip: Each panda is connected to a different piece of fruit!

47

Finish

48

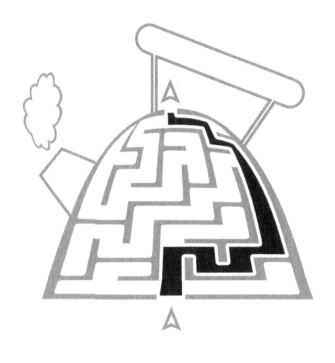

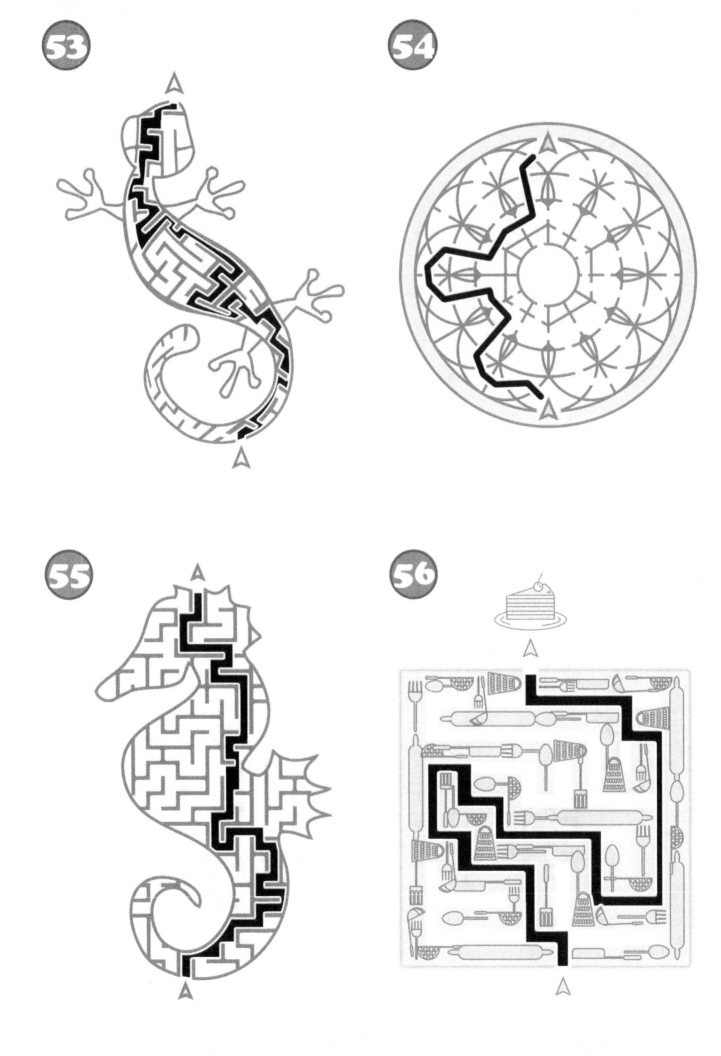

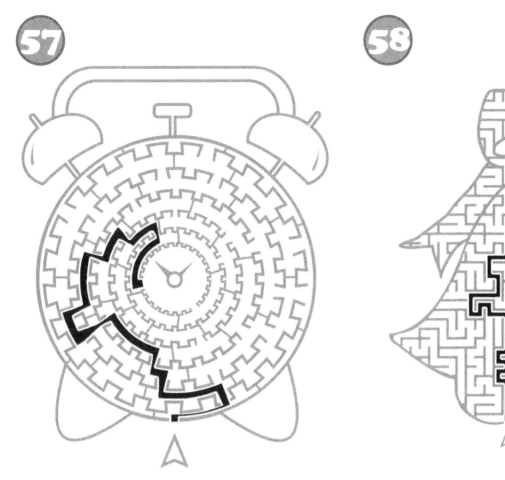

Tip: Try to get to the center of the clock!

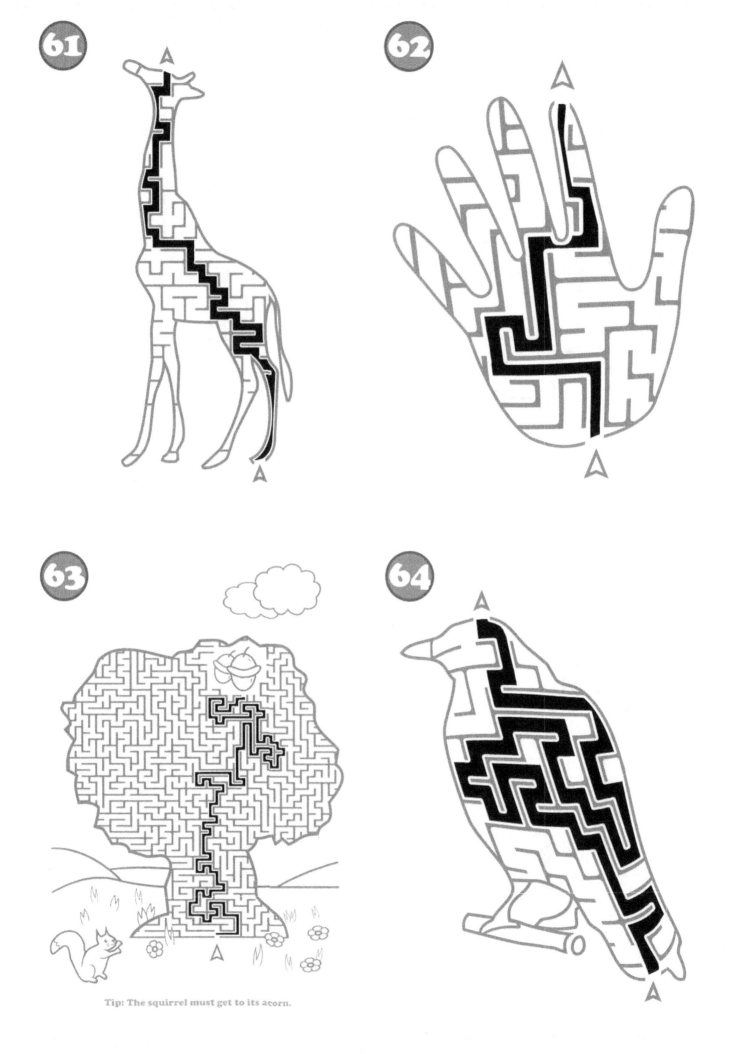

61

62

63

Tip: The squirrel must get to its acorn.

64

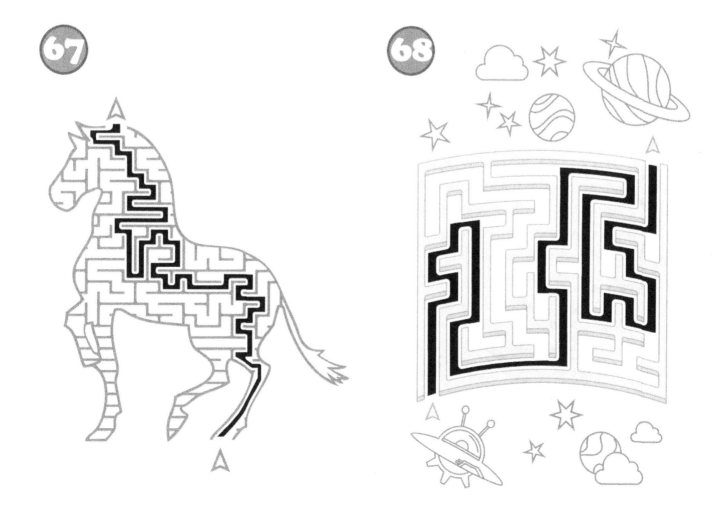

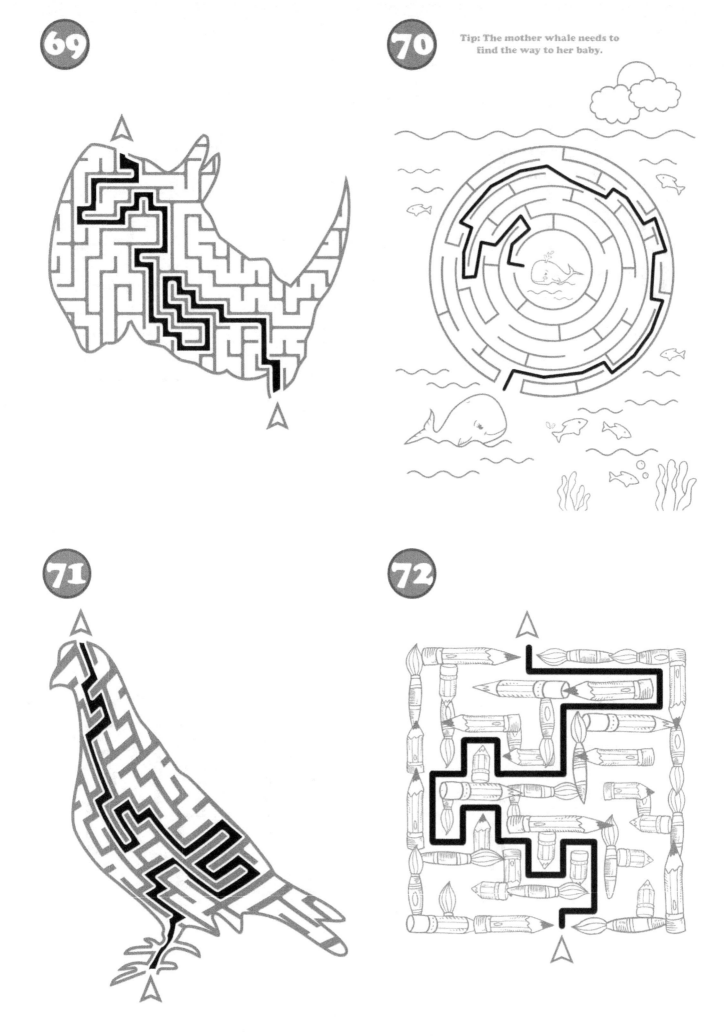

69

70 Tip: The mother whale needs to find the way to her baby.

71

72

Tip: The gingerbread man needs to
find the way home.

Tip: Connect each animal to their tracks.

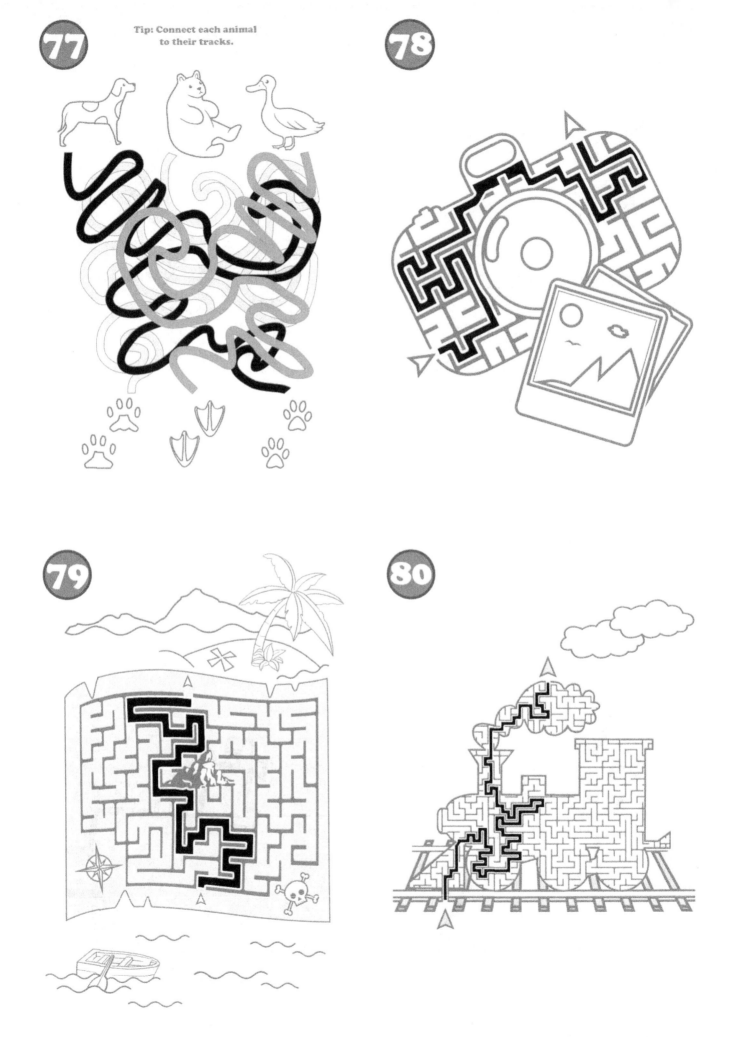

Made in the USA
Middletown, DE
25 October 2022

13517907R00057